Lecture Notes in Mathematics

An informal series of special lectures, seminars and reports on mathematical topics

Edited by A. Dold, Heidelberg and B. Eckmann, Zürich

19

Gabriel Stolzenberg

Brown University
Providence, Rhode Island

Volumes, Limits, and Extensions of Analytic Varieties

1966

Springer-Verlag · Berlin · Heidelberg · New York

Library of Congress Catalog Card Number 66-25792. Printed in Germany. Title No. 7339.

Contents

Introduction

This exposition is based on lectures I gave in November 1965 at the Brown Analysis Seminar. My main source was E. Bishop's " Conditions for the analyticity of certain sets" [1] . My other principal references were H. Federer's " Some theorems on integral currents" [2] and H. Rutishauser's " Über Folgen und Scharen von analytischen und meromorphen Funktionen mehrerer Variablen, sowie von analytischen Abbildungen" [7] .

Here are the theorems that will be covered.

(A) (Wirtinger) An analytic subvariety of a domain in $\mathbb{C}^n$ minimizes volume (globally) .

(B) If a purely k-dimensional subvariety of an R-ball in $\mathbb{C}^n$ passes through the center of the ball then its 2k-volume is at least $\alpha(2k)\,R^{2k}$ (where , for any positive integer d , $\alpha(d)$ is the volume of the unit ball in Euclidean d-space.)

(C) (Bishop) The limit of a sequence of purely k-dimensional analytic varieties whose 2k-volumes are uniformly bounded is again a purely k-dimensional variety.

(D) (Stoll) A purely k-dimensional global subvariety of $\mathbb{C}^n$ whose intersection with every R-ball about O has 2k-volume $O(R^{2k})$ is algebraic.

These theorems will be proved in Chapters I-IV. Then in the Appendix I will discuss briefly two more results of Bishop which are closely related to (C) and (D). Namely,

(E) (Bishop) Let W be a subvariety of a domain U. If V is a purely k-dimensional subvariety of U-W such that $\bar{V} \cap W$ (closure in U) has zero 2k-dimensional Hausdorff measure then $\bar{V}$ is an analytic subvariety of U.

(F) (Bishop) Let W be a subvariety of a domain U. If V is a purely k-dimensional subvariety of U-W whose 2k-volume is finite then $\bar{V}$ is an analytic subvariety of U.

For nonsingular varieties (A) is Wirtinger's Theorem [11]. For general varieties it is proved by Federer in [2]. He formulates Theorem (A), most naturally, in terms of currents, and also gives a very simple proof of the basic Wirtinger Inequality. I will present his approach here, but without currents.

Federer also showed me how to derive (B) from (A) , by a technique which has been used in recent years in work on the Plateau problem . I do not know its origin. (Briefly, the method is this. Compare the volume of the variety with that of the cone through O over the intersection of the variety with the r-sphere about O , for each $r < R$. Then integrate.)

Theorems (C), (E) and (F) were all proved by Bishop in [1] . I will present his proof of (C) in Chapters III and IV ; and in the Appendix I will discuss the reduction of (F) to (E) .

Theorem (D) is Stoll's criterion for a global subvariety of $\mathbb{C}^n$ to be algebraic [8] . (It is also necessary.) In Chapter IV I will derive (D) from (C) by following an old idea of Rutishauser [7] . This yields the simplest proof of Stoll's criterion. (And Chow's Theorem now appears, where it belongs, as a corollary.)

Theorem (D) is also an immediate consequence of (F) . (Essentially, take $U = \mathbb{CP}^n$ complex projective n-space and W = the hyperplane at infinity. The growth condition on the volume of $V \subset \mathbb{C}^n$ in (D) implies that V has finite volume when viewed in $\mathbb{CP}^n = \mathbb{C}^n \cup W$ with the standard Kähler metric on projective space. Then (F) gives $\overline{V}$ as a subvariety of $\mathbb{CP}^n$ - and Chow's Theorem, which is also a corollary of (F) , says that V is algebraic.)

The earliest statement of (C) that I have seen - for curves in $\mathbb{C}^2$

is that of K. Oka in his pre-Roman numerals 1934 note [5]. In 1950 in [7] Rutishauser presented a controversial proof of this case. It did not win general acceptance, probably because of his unconvincing statement " Nach § I, h) kann jetzt J noch aus analytischen Flächen oder analytischen Hyperflächen bestehen" on p. 264, on which his argument apparently depended. In this same paper Rutishauser showed how to get (D) from (C) for curves in $\mathbb{C}^2$. In Chapter IV I will extend this to the general case.

Finally, in 1962, Theorem (C) was proved for curves in $\mathbb{C}^2$ by Nishino [4] and Oka [6]; and in 1964, by a somewhat similar method, Stoll [9] settled it for hypersurfaces. In this approach the volume of a hypersurface $\{ f = 0 \}$ in a domain U is related to the integral $\int_U \log |f| \, dU$.

Rutishauser's attack on (C) was different. He relied on a local lower estimate like (B) to be played off against the given upper bound. This is the same approach that Bishop [1] followed with great success in (C) and (F) - and which I will present in Chapters III and IV. (Bishop's proof of (C) shows, incidentally, that Rutishauser's method works.)

The reduction of (F) to (E) also requires a local lower bound. Namely, *suppose that* B *is an* R-*ball in* $\mathbb{C}^n$ *and* V *is a purely* k-*dimensional subvariety of* $B - \{ z_n = 0 \}$ *whose closure* $\overline{V}$ *passes through the center of* B . *Then, for some constant* c(k,n) , *independent of* V *and* R ,

$$(*) \qquad \mathrm{Vol}_{2k}(V) \geq c(k,n) \cdot R^{2k} .$$

This estimate can also be used in place of (B) to prove (C) , but (B) gives the best value of the constant, $c(k,n) = \alpha(2k)$. In [1] Bishop worked out this local lower estimate only for $k = 1$ and asserted that the general case would follow by induction on k . A more detailed sketch of his argument will be presented in the Appendix.

(My own interest in these problems stems from the time [10] I constructed a sequence of analytic curves in a ball in $\mathbb{C}^2$, all passing through the center, such that the limit set did not even *contain* any analytic disks. The areas grew (inevitably) tremendously, and the construction contradicted Rutishauser's assertion about "J" on p. 264 of [7] .

I believe all these matters are now firmly settled.)

All the background material on analytic varieties that will be needed is contained in the first three chapters of " Analytic Functions of Several Complex Variables" by R. C. Gunning and H. Rossi [3] .

CHAPTER I

Analytic Varieties Minimize Volume

The exact statement will come later. (It is complicated by the possible presence of singularities.) The proof has two ingredients, Wirtinger's Inequality and Stokes' Theorem.

Wirtinger's Inequality. Let L be a complex linear space and let M be a real even-dimensional subspace. Let H be a positive definite Hermetian form on L . Then $H = S + iA$ where S is symmetric and A is alternating. Let $\{ m_1, \ldots, m_{2k}\}$ be a basis of M which is orthonormal with respect to S . Then

$$|A^k(m_1, \ldots, m_{2k})| \leq k\,!$$

with equality holding precisely when M is a complex k-dimensional subspace of L . (Here A^k is the k-th exterior power of A .)

Proof: ([2]) Firstly, the value of $|A^k(m_1, \ldots, m_{2k})|$ does not depend on the choice of orthonormal basis of M . Secondly, the restriction of A to M , A_M , has a certain canonical representation. Namely, there is an orthonormal basis $m'_1, \ldots, m'_{2k}$

of M with dual basis $\varphi_1, \ldots, \varphi_{2k}$ such that

$$A_M = \sum_{j=1}^{k} a_j(\varphi_{2j-1} \wedge \varphi_{2j})$$

where $a_j = A(m'_{2j-1}, m'_{2j})$. Hence, on M,

$$A^k = k! \bigwedge_{j=1}^{k} a_j(\varphi_{2j-1} \wedge \varphi_{2j}),$$

so that

$$|A^k(m'_1, \ldots, m'_{2k})| = k! \prod_{j=1}^{k} |a_j| .$$

Therefore, the general result will follow directly from the case $k = 1$ applied to the real span of each pair $\{ m'_{2j-1}, m'_{2j} \}$.

But when $k = 1$ we have $A(m_1, m_2) = -iH(m_1, m_2)$ and $H(m_1, m_1) = 1 = H(m_2, m_2)$ (because $S(m_\mu, m_\nu) = \delta_{\mu\nu}$ and A is alternating.) Therefore, by Schwarz's Inequality applied to H, $|A(m_1, m_2)| \leq 1$ with equality if and only if $m_2 = cm_1$ for some <u>complex</u> number $c \neq 0$.

Q. E. D.

<u>Note.</u> By a suitable shuffling of the basis it can be arranged that $A^k(m_1, \ldots, m_{2k}) \geq 0$.

To apply Wirtinger's Inequality let $z_1, \ldots, z_n$ be coordinates on complex n-space $\mathbb{C}^n$ and set

$$\Omega = \frac{i}{2} \sum_{j=1}^{n} dz_j \wedge d\bar{z}_j \ .$$

This is the fundamental 2-form of the standard Kähler structure on $\mathbb{C}^n$ and, at each point $p \in \mathbb{C}^n$, Ω_p is the alternating part of the positive definite Hermetian form

$$\sum_{j=1}^{n} dz_{j(p)} \cdot d\bar{z}_{j(p)}$$

on the tangent space to $\mathbb{C}^n$ at p. Therefore, if $\mathfrak{M}$ is any smooth 2k-dimensional manifold immersed in $\mathbb{C}^n$, Wirtinger's Inequality implies immediately that

$$\int_{\mathfrak{M}} \frac{1}{k!} \Omega^k \leq \int_{\mathfrak{M}} 1 \, d\mathfrak{M} = \text{Volume}_{2k}(\mathfrak{M})$$

with equality if and only if $\mathfrak{M}$ is a complex k-dimensional manifold. Also, each $\frac{1}{k!}\Omega^k$ is an exact 2k-form. $\frac{1}{k!}\Omega^k = d\,\Phi_k$ for some 2k-1 form Φ_k which can be computed. Hence, by Stokes' Theorem we can now deduce the non-singular version of Theorem(A). Namely,

Let $(\mathfrak{M}, \partial)$ and $(\mathfrak{N}, \partial)$ be smooth compact 2k-manifolds with boundary immersed in $\mathbb{C}^n$ and having the same boundary ∂. If $\mathfrak{M}$ is a complex k-manifold then $\mathrm{Vol}_{2k}(\mathfrak{M}) \leq \mathrm{Vol}_{2k}(\mathfrak{N})$.

Proof: $\mathrm{Vol}_{2k}(\mathfrak{M}) = \int_{\mathfrak{M}} \frac{1}{k!}\Omega^k = \int_{\partial}\Phi_k = \int_{\mathfrak{N}} \frac{1}{k!}\Omega^k \leq \mathrm{Vol}_{2k}(\mathfrak{N})$.

Now I will state the more general version of (A) that will be proved here.

THEOREM (A). Let W be a purely k-dimensional analytic subvariety of a domain in $\mathbb{C}^n$ and let S be its singular locus. Let V be an open relatively compact subset of W, with topological boundary ∂ in W, and such that $(V - S\cap V, \partial - S\cap\partial)$ is a smooth 2k-manifold with boundary and $\partial - S\cap\partial$ has finite 2k-1 volume. Let $\mathfrak{M}_V = V - S\cap V$ and $\partial_V = \partial - S\cap\partial$. Let $\mathfrak{N}$ be any smooth relatively compact 2k-manifold in $\mathbb{C}^n$ with the properties that $(\mathfrak{N}, \partial_V)$ is a (not necessarily compact) manifold with boundary and that the set $\delta = \overline{\mathfrak{N}} - (\mathfrak{N}\cup\partial_V)$ (closure in $\mathbb{C}^n$) is a compact countable union of smooth manifolds of dimensions $\leq 2k-2$. Then $\mathrm{Vol}_{2k}(\mathfrak{M}_V) \leq \mathrm{Vol}_{2k}(\mathfrak{N})$.

(Note. The inequality is strict unless $\mathfrak{N}$ is an open dense subset of $\mathfrak{M}_V$ but that will not be proved here.)

The proof is basically the same as in the non-singular case. It is only a matter of justifying the step

$$\int_{\mathcal{M}_V} \frac{1}{k!}\,\Omega^k = \int_{\partial_V} \Phi_k = \int_{\eta} \frac{1}{k!}\,\Omega^k .$$

This will follow directly from the fact that $\mathrm{Vol}_{2k}(\mathcal{M}_V) < \infty$ (to be proved) and the following extension of Stokes' Theorem.

Stokes Extended. Suppose $\mathcal{M}$ *is a smooth relatively compact d-manifold in* $\mathbb{R}^e$ *with finite d-volume and such that* $\bar{\mathcal{M}} - \mathcal{M}$ *(closure in* $\mathbb{R}^e$ *) can be decomposed into* $\partial \cup \delta$ *where*

(i) $(\mathcal{M}, \partial)$ *is a (not necessarily compact) manifold-with-boundary*

(ii) ∂ *has finite* d-1 *volume*

(iii) δ *is compact*

(iv) the image of δ under each coordinate projection $(x_1, \ldots, x_e) \longrightarrow (x_{i_1}, \ldots, x_{i_{d-1}})$ has measure 0 in $\mathbb{R}^{d-1}$.

Then, for any d-1 form τ on $\mathbb{R}^e$,

$$\int_{\partial} \tau = \int_{\mathcal{M}} d\tau .$$

Proof: ([2]) For each $I = (i_1, \ldots, i_{d-1})$ (where

$1 \leq i_1 < \ldots < i_{d-1} \leq e$) set $\omega_I = dx_{i_1} \wedge \ldots \wedge dx_{i_{d-1}}$ and let $\Pi_I : \mathbb{R}^e \longrightarrow \mathbb{R}^{d-1}$ be the associated projection $(x_1, \ldots, x_e) \longrightarrow (x_{i_1}, \ldots, x_{i_{d-1}})$. Also define $\partial_I = \partial - \Pi_I^{-1}(\Pi_I(\delta))$ and $\mathcal{M}_I = \mathcal{M} - \Pi_I^{-1}(\Pi_I(\delta))$. The restrictions on δ imply that for any smooth function f and any positive integer $j \leq e$

$$\int_\partial f\omega_I = \int_{\partial_I} f\omega_I$$

and

$$\int_{\mathcal{M}} f\,\omega_I \wedge dx_j = \int_{\mathcal{M}_I} f\,\omega_I \wedge dx_j .$$

For each I choose a sequence of smooth functions Ψ_m^I on $\mathbb{R}^{d-1}$ such that $0 \leq \Psi_m^I \leq 1$, $\Psi_m^I \equiv 0$ on a neighborhood of $\Pi_I(\delta)$, and $\Psi_m^I \longrightarrow 1$ uniformly on compact subsets of $\mathbb{R}^{d-1} - \Pi_I(\delta)$.

Then set $\Phi_m^I = \Psi_m^I \circ \Pi_I$, express τ as $\Sigma \tau_I \omega_I$, and define

$$\tau_m = \Sigma \tau_I \Phi_m^I \omega_I .$$

Since these forms τ_m vanish identically on (varying) neighborhoods of δ, the standard version of Stokes' Theorem applies -

$$\int_{\mathcal{M}} d\tau_m = \int_\partial \tau_m .$$

Now computing

$$\int_{\partial} \tau_m = \sum \int_{\partial} \tau_I \Phi^I_m \omega_I = \sum \int_{\partial_I} \tau_I \Phi^I_m \omega_I \longrightarrow$$

$$\sum \int_{\partial_I} \tau_I \omega_I = \int_{\partial} \tau \quad .$$

Also, $d\tau_m = \Sigma \Phi^I_m d(\tau_I \omega_I)$ because each $d\Phi^I_m \wedge \omega_I \equiv 0$.
Therefore, if we express $d(\tau_I \omega_I)$ as

$$\sum_{j=1}^{e} \sigma_{I,j} \, \omega_I \wedge dx_j$$

then

$$\int_{\mathcal{M}} d\tau_m = \sum_{j=1}^{e} \sum_{I} \int_{\mathcal{M}_I} \Phi^I_m \sigma_{I,j} \, \omega_I \wedge dx_j \longrightarrow$$

$$\sum_{j=1}^{e} \sum_{I} \int_{\mathcal{M}_I} \sigma_{I,j} \, \omega_I \wedge dx_j = \int_{\mathcal{M}} d\tau \quad .$$

Hence

$$\int_{\partial} \tau = \int_{\mathcal{M}} d\tau \; .$$

Q.E.D.

Note. The finite volume assumptions were used in taking limits.

What needs to be checked in order to apply this extension of

Stokes' Theorem to $\mathcal{N}, \partial_V$ and $\mathcal{M}_V, \partial_V$? Conditions (i)-(iii) are evident for $\mathcal{N}$, ∂_V and $\delta = \overline{\mathcal{N}} - (\mathcal{N} \cup \partial_V)$; and (iv) obtains because δ is a countable union of smooth manifolds of dimensions $\leq 2k-2$. Therefore, if $\mathrm{Vol}_{2k}(\mathcal{N}) < \infty$ then the extension applies and

$$\int_{\mathcal{N}} \frac{1}{k!} \Omega^k = \int_{\partial_V} \Phi_k \ .$$

On the other hand, when $\mathrm{Vol}_{2k}(\mathcal{N}) = \infty$ the conclusion of (A) is immediate .

Similarly, (i)-(iii) are direct for $\mathcal{M}_V$, ∂_V and $\overline{V} \cap S = \overline{\mathcal{M}}_V - (\mathcal{M}_V \cup \partial_V)$. Condition (iv) holds because $\overline{V} \cap S$ is contained in the singular locus S which, being an analytic variety of dimension $\leq k-1$, is (by induction) a countable union of complex manifolds of dimensions $\leq k-1$. Therefore, the only missing piece is

$$\mathrm{Vol}_{2k}(\mathcal{M}_V) < \infty \ .$$

Since $\overline{\mathcal{M}}_V = \overline{V}$ is a compact subset of the variety W it suffices to prove the following local result.

LEMMA . For each point $p \in W$ there is a neighborhood N such that $\mathrm{Vol}_{2k}(N \cap (W - S)) < \infty$.

Proof: We may take p to be the origin O . (The only problem is, of course, when $O \in S$ the singular locus.)

Since W is a purely k-dimensional variety, for a suitable choice of coordinates $z_1, \ldots, z_n$ on $\mathbb{C}^n$, each (n-k)-plane of the form $\{ z_{i_1} = 0, \ldots, z_{i_k} = 0 : 1 \leq i_1 < \ldots < i_k \leq n \}$ will meet W (near O) only at O . (Almost all coordinate systems have this property.) Then, by the standard local analysis of a variety (see Chapter III of [3] , especially Corollary 9 of Section C) there is a polydisk D about O in $\mathbb{C}^k$, and for each $I = (i_1, \ldots, i_k)$ a neighborhood N_I of O in $\mathbb{C}^n$, such that the projection $\Pi_I : \mathbb{C}^n \longrightarrow \mathbb{C}^k$, $(z_1, \ldots, z_n) \longrightarrow (z_{i_1}, \ldots, z_{i_k})$ maps $N_I \cap W$ as a finite-sheeted branched covering of D . Let s(I) be the sheet number. Also, for each I there is a hypersurface Δ_I of D such that $N_I \cap S \subset \Pi_I^{-1} (\Delta_I)$ and

$$N_I \cap W - \Pi_I^{-1}(\Delta_I) \xrightarrow{\Pi_I} D - \Delta_I$$

is a regular s(I)-sheeted covering.

Let N be the intersection of the N_I and let T be the union of the $\Pi_I^{-1}(\Delta_I)$ in W . Then T is a hypersurface in W , so its intersection with the 2k-manifold W - S has measure 0 . Therefore

$$\mathrm{Vol}_{2k}(N \cap (W-S)) = \int_{N \cap (W-T)} \frac{1}{k!}\,\Omega^k \leq \sum s(I) \cdot \mathrm{Vol}_{2k}(D) .$$

Q. E. D.

This justifies

$$\int_{\mathfrak{M}_V} \frac{1}{k!}\,\Omega^k = \int_{\partial V} \Phi_k$$

and completes the proof of Theorem (A) .

(Note . The most general and natural version of (A) says that an open relatively compact subset of an analytic variety is a minimal current [2].)

CHAPTER II

A Local Lower Bound for the Volume of an Analytic Variety

If V is a purely k-dimensional subvariety of a domain in $\mathbb{C}^n$ define $Vol_{2k}(V) = Vol_{2k}(V-S)$. (Note that $V-S$ is a smooth $2k$-manifold and S is a countable union of smooth manifolds of lower dimension.)

For any Euclidean space $\mathbb{R}^d$ let $\|p\|$ denote the distance from a point p to O and let $\alpha(d)$ be the volume of the unit ball. For each $r > 0$ let $B(p;r)$ be the open ball of radius r about the point p.

THEOREM (B) Let $R > 0$ and let V be a purely k-dimensional subvariety of $B(O;R)$ in $\mathbb{C}^n$ such that $O \in V$. Then $Vol_{2k}(V) \geq \alpha(2k) \cdot R^{2k}$.

Proof: Two propositions will be proved.

PROPOSITION 1. For each positive $r < R$ let $V_r = V \cap B(O;r)$. Then

$$\frac{1}{r^{2k}} \cdot Vol_{2k}(V_r)$$

is (weakly) increasing.

PROPOSITION 2. <u>Let</u> $d \leq e$ <u>be positive integers.</u> <u>Let</u> M <u>be a smooth</u> d-<u>dimensional submanifold of a neighborhood of</u> O <u>in</u> $\mathbb{R}^e$ <u>such that</u> $O \in M$, <u>and define</u> $M_r = \{ m \in M : \|m\| < r \}$. <u>Then</u>

$$\lim_{r \to 0} \frac{1}{\alpha(d) r^d} \cdot \underline{\mathrm{Vol}}_d(M_r) = 1 .$$

From these two propositions it follows immediately that Theorem (B) is true whenever $O \in V-S$ (where S is the singular locus of V .) But even if $O \in S$ there is a sequence v_i in $V-S$ with $v_i \to O$, and a sequence $r_i \nearrow R$ such that $B(v_i ; r_i) \subset B(O ; R)$. Then $\mathrm{Vol}_{2k}(V \cap B(v_i ; r_i)) \longrightarrow \mathrm{Vol}_{2k}(V)$, so Theorem (B) for each $V \cap B(v_i ; r_i)$ implies also $\mathrm{Vol}_{2k}(V) \geq \alpha(2k) \cdot R^{2k}$. Therefore, it suffices to prove the two propositions.

<u>Proof of 2.</u> (Notice that all we need for (B) is that the $\underline{\lim}$ is ≥ 1 .)

Let T be the tangent space to M at O and choose coordinates $x_1, \ldots, x_e$ on $\mathbb{R}^e$ so that $T = \{ x_{d+1} = 0 , \ldots , x_e = 0 \}$. Then, near O , M is defined by equations $x_j = h_j(x_1, \ldots, x_d)$ $j = d+1, \ldots, e$ where $h_j(O) = 0$ and $dh_j(O) = 0$. Write $x = (x_1, \ldots, x_d)$ and $h(x) = (h_{d+1}(x), \ldots, h_e(x))$. Then $\|h(x)\| = o(\|x\|)$. Therefore, if we define

$$m(r) = \min \{ \|x\| : \|(x, h(x))\| = r \}$$

then $m(r)/r \longrightarrow 1$ as $r \longrightarrow 0$. To use this information let $\Pi: \mathbb{R}^e \longrightarrow T$ be the projection $(x_1, \ldots, x_e) \longrightarrow (x_1, \ldots, x_d, 0, \ldots, 0)$. For each $s > 0$ let $T_s = \{ p \in T : \|p\| < s \}$. Then $\mathrm{Vol}_d(T_s) = \alpha(d) \cdot s^d$ and, for all small r,

$$T_{m(r)} \subset \Pi(M_r) \subset T_r \ .$$

Hence, $\alpha(d).(m(r))^d \leq \mathrm{Vol}_d(\Pi(M_r)) \leq \alpha(d) \cdot r^d$, so

$$\lim_{r \to 0} \frac{1}{\alpha(d) r^d} \cdot \mathrm{Vol}_d(\Pi(M_r)) \longrightarrow 1 \ .$$

But $\mathrm{Vol}_d(M_r) \geq \mathrm{Vol}_d(\Pi(M_r))$ so that

$$\underline{\lim}_{r \to 0} \frac{1}{\alpha(d) r^d} \cdot \mathrm{Vol}_d(M_r) \geq 1 \ .$$

To complete the proof of Proposition 2 view T as $\mathbb{R}^d$ with coordinates $x_1, \ldots, x_d$ and set $U_r = \Pi(M_r)$. Then, by linear algebra, if r is small enough, an upper bound for $\mathrm{Vol}_d(M_r)$ is given by the integral over U_r of the sum of the absolute values of the determinants of the $d \times d$ submatrices of the Jacobian matrix of the mapping

$(x_1, \ldots, x_d) \longrightarrow (x_1, \ldots, x_d, h_{d+1}(x), \ldots, h_e(x))$. One summand integrates out to $\mathrm{Vol}_d(U_r)$; and every other one involves the gradient of at least one h_j , so that its integral over U_r is $o(\mathrm{Vol}_d(U_r))$. But $U_r = \Pi(M_r)$ so we have

$$\lim_{r \to 0} \frac{1}{\alpha(d) r^d} \cdot \mathrm{Vol}_d(M_r) =$$

$$\lim_{r \to 0} \frac{1}{\alpha(d) r^d} \cdot \mathrm{Vol}_d(\Pi(M_r)) = 1 .$$

Proof of 1 . This will be an application of (A) . Let $F(r) = \mathrm{Vol}_{2k}(V_r)$ for $r < R$. This is a monotone function of r . By integrating and exponentiating, to prove 1 it is enough to show that

$$\frac{2k}{r} \leq \frac{F'(r)}{F(r)} \qquad \text{a.e. } dr \ ;$$

which I will rewrite as

$$F(r) \leq \frac{r}{2k} \cdot F'(r) \quad \text{a.e. } dr$$

so as to express an inequality between volumes. I claim that the expression on the right is the volume of the open cone $\mathcal{C}_r$ with vertex O over $\partial_r = \{ p \in V-S : \|p\| = r \}$, and that (for almost all r) Theorem (A) applies, yielding the desired inequality.

To justify this, consider the map $p \longrightarrow \|p\|^2$ on $V-S$. It is real analytic so, except for isolated values of r, ∂_r is a real analytic manifold. Then we can express $\mathrm{Vol}_{2k}(V_r)$ as $\int_o^r \mathrm{Vol}_{2k-1}(\partial_t)dt$, so that

$$F'(r) = \mathrm{Vol}_{2k-1}(\partial_r) < \infty \qquad \text{a.e. } dr .$$

Therefore, $\frac{r}{2k} F'(r)$ is indeed the volume of the cone $\mathcal{C}_r$.

To be able to apply Theorem (A) it remains only to examine $\delta_r = \bar{\mathcal{C}}_r - (\mathcal{C}_r \cup \partial_r)$. (Everything else is in order.) But δ_r is the closed cone over $S \cap \{p : \|p\| = r\}$, so it is compact. Also, since S is itself a variety of dimension $\leq k-1$ we have, by the same reasoning as above, that (for almost all r) $S \cap \{p : \|p\| = r\}$ is a finite union of real analytic manifolds of dimensions $\leq 2k-3$. Therefore, the cone δ_r is contained in a finite union of smooth manifolds of dimensions $\leq 2k-2$, so that Theorem (A) applies and we are done.

CHAPTER III

Hausdorff Measure and the Hausdorff Metric

The metric will be used to take a limit of analytic varieties. The measure is in lieu of a volume element on the limit set.

DEFINITION. Let X be a metric space and let d be any non-negative real number. Let S be any subset of X. For each $\varepsilon > 0$ let $I(S, \varepsilon)$ be the infimum of all sums of the form

$$\sum_{j=1}^{\infty} (\operatorname{diam} S_j)^d$$

where $S = \bigcup_{j=1}^{\infty} S_j$ and each $\operatorname{diam} S_j < \varepsilon$. $I(S, \varepsilon)$ increases (weakly) as $\varepsilon \searrow 0$, and we define _the Hausdorff_ d-_measure of_ S to be

$$H_d(S) = \frac{1}{2^d} \cdot \lim_{\varepsilon \to 0} I(S, \varepsilon).$$

(The $\frac{1}{2^d}$ is purely for aesthetic reasons.)

Here are some important elementary properties.

PROPERTY 1. *If* $H_d(S) < \infty$ *and* $d < e$ *then* $H_e(S) = 0$.

PROPERTY 2. *Suppose* f *is a Lipschitz map from* X *to another metric space* Y *with Lipschitz constant* K. (*For all* $x_1, x_2 \in X$ $\operatorname{dist}(f(x_1), f(x_2)) \leq K \operatorname{dist}(x_1, x_2)$.) *Then, for any* $S \subset X$ *and* $d > 0$, $H_d(f(S)) \leq K^d \cdot H_d(S)$. *In particular, if* $H_d(S) = 0$ *then* $H_d(f(S)) = 0$.

PROPERTY 3. *If* $X = \mathbb{R}^e$ *and* S *is a smooth* d-*manifold in* $\mathbb{R}^e$ *then* $\operatorname{Vol}_d(S) = \alpha(d) \cdot H_d(S)$.

The first two properties are direct consequences of the definitions. The third can be quickly verified by means of Property 2 and the reasoning used to prove Proposition 2. I will not carry out the argument here. (But I remark that all we will need is that there are *some* positive constants A_d, B_d such that $A_d \cdot H_d(S) \leq \operatorname{Vol}_d(S) \leq B_d \cdot H_d(S)$.)

Next I will define the Hausdorff metric.

DEFINITION. Let K be a compact metric space and let $\operatorname{Comp}(K)$ be the set of all compact subsets of K. *The Hausdorff metric on* $\operatorname{Comp}(K)$ is defined by

$$\underline{\text{dist}}(S, T) = \max_{s \in S}\left\{\min_{t \in T} \text{dist}(s,t)\right\} + \max_{t \in T}\left\{\min_{s \in S} \text{dist}(t,s)\right\}.$$

This is indeed a metric and with it $\underline{\text{Comp}}(K)$ is itself a compact space.

For any metric space X, if S_i $i = 1, 2, \ldots$ and S are closed subsets I will say that

$$S_i \longrightarrow S$$

provided that for every compact $K \subset X$, $S_i \cap K$ is a convergent sequence in $\underline{\text{Comp}}(K)$ and

$$S = \bigcup_K \lim_{i \to \infty} (S_i \cap K).$$

Note that it does not follow that $S_i \cap K \longrightarrow S \cap K$ for every compact K. (For example, if $p \in S$ but p is not in any S_i then $S_i \cap \{p\} = \emptyset \not\rightarrow S \cap \{p\}$.)

If X can be expressed as a countable union of compact sets then, by a diagonal process, every sequence S_i yields a convergent subsequence. This is the sense in which we will discuss limits of analytic subvarieties of a domain in $\mathbb{C}^n$.

The next proposition relates the metric and the measure, and shows how a local lower bound like that of Theorem (B) can be played off against an upper bound.

PROPOSITION 3. (Bishop) *Let* X *be a metric space and* $d \geq 0$. *Let* $S_i \longrightarrow S$ *as closed subsets of* X *and suppose there is a constant* $M > 0$ *such that every* $H_d(S_i) < M$. *Suppose also there is a constant* $N >$ *and, for each compact* $K \subset X$ *an* $r(K) > 0$ *such that, for* p *in any* $S_i \cap K$ *and* $r \leq r(K)$, $H_d(S_i \cap B(p; r)) \geq N \cdot r^d$ (*where* $B(p; r)$ *is the open metric ball of radius* r *about* p .) *Then, for any compact* $K \subset X$,

$$H_d (\lim_{i \to \infty} (S_i \cap K)) \leq 4^d M/N .$$

If X *is* σ*-compact then* $H_d(S) \leq 4^d M/N.$

Proof: Let $S_K = \lim_{i \to \infty} (S_i \cap K)$. If $p \in S_K$ there is a sequence $p_i \in S_i \cap K$ with $p_i \longrightarrow p$. By the assumed local lower bound it must be that for $\varepsilon \leq r(K)$

$$\underline{\lim}_{i \to \infty} H_d(S_i \cap B(p; \varepsilon)) \geq N \cdot \varepsilon^d .$$

Also, for any $\varepsilon > 0$ there are finitely many points $p_1(\varepsilon), \ldots, p_{n(\varepsilon)}(\varepsilon)$ *in* S_K such that the balls $B(p_i(\varepsilon) ; \varepsilon/2)$ are all disjoint and

$$S_K \subset \bigcup_i B(p_i(\varepsilon) ; \varepsilon) .$$

This can be demonstrated as follows. Since S_K is compact and is covered by $\{ B(p; \epsilon/2) : p \in S_K \}$ there is a finite subcover- say $B(q_1 ; \epsilon/2), \ldots, B(q_m; \epsilon/2)$. Let $p_1(\epsilon) = q_1$. Remove all $B(q_j; \epsilon/2)$ which meet $B(p_1(\epsilon); \epsilon/2)$. They all lie in $B(p_1(\epsilon); \epsilon)$. Let $B(q_{j_2} ; \epsilon/2)$ be the next ball that remains. Let $p_2(\epsilon) = q_{j_2}$ and continue.

Combining these remarks, we now have, for $0 < \epsilon \leq r(K)$,

$$M \geq \overline{\lim_{i \to \infty}} H_d(S_i)$$

$$\geq \overline{\lim_{i \to \infty}} \sum_{j=1}^{n(\epsilon)} H_d (S_i \cap B(p_j(\epsilon) ; \epsilon/2))$$

$$\geq \sum_{j=1}^{n(\epsilon)} N(\epsilon/2)^d \quad ;$$

and, from the definition of Hausdorff measure,

$$H_d(S_K) \leq \lim_{\epsilon \to 0} \sum_{j=1}^{n(\epsilon)} (2\epsilon)^d \quad .$$

Therefore, $H_d(S_K) \leq 4^d M/N$.

If X is σ - compact express it as the union of a sequence of compact subsets K_t $t = 1, 2, \ldots$ and let $r_t = r(K_t)$. Let ϵ_t be a decreasing sequence of positive numbers such that $\epsilon_t \leq r_t$. For each S_{K_t} we have associated points $p_1(\epsilon_t), \ldots, p_{n(\epsilon_t)}(\epsilon_t)$. If we arrange these points in the order

$$p_1(\varepsilon_1),\ldots,p_{n(\varepsilon_1)}(\varepsilon_1),\ p_1(\varepsilon_2),\ldots,p_{n(\varepsilon_2)}(\varepsilon_2),\ldots$$

and repeat the elimination proceedure used above we obtain a sequence of points

$$x_1(\varepsilon_1),\ldots,x_{m(\varepsilon_1)}(\varepsilon_1),\ x_1(\varepsilon_2),\ldots,x_{m(\varepsilon_2)}(\varepsilon_2),\ldots$$

such that the balls $B(x_i(\varepsilon_t)\,;\,\varepsilon_t/2)$ are <u>all</u> disjoint and

$$S = \bigcup_t S_{K_t} \subset \bigcup_{i,t} B(x_i(\varepsilon_t)\,;\,\varepsilon_t)\ .$$

Then, as above, we get

$$M \geq \sum_{i,t} N\cdot(\varepsilon_t/2)^d \qquad \text{and} \quad H_d(S) \leq \lim_{\varepsilon_1\to 0} \sum_{i,t}(2\varepsilon_t)^d\ ;$$

so that $H_d(S) \leq 4^d M/N$.

APPLICATION . *Let* U *be a domain in* $\mathbb{C}^n$ *and let* V_i *be a sequence of purely* k-*dimensional subvarieties of* U *which converges to* V *some closed subset of* U . *If the* $\mathrm{Vol}_{2k}(V_i)$ *are uniformly bounded above then* $H_{2k+1}(V) = 0$.

For $\mathrm{Vol}_{2k}(V_i) = \mathrm{Vol}_{2k}(V_i - S_i)$ where S_i is the singular locus of V_i ; and it follows directly from the subadditivity of Hausdorff measure and from Property 3 that $\mathrm{Vol}_{2k}(V_i) = \alpha(2k) \cdot H_{2k}(V_i)$ even when there are singularities present. Therefore, by Proposition 3, $H_{2k}(V) < \infty$; so by Property 1 $H_{2k+1}(V) = 0$.

This can be used in the following way to get a certain valuable proper mapping.

PROPOSITION 4. (Bishop) Let U be a domain in $\mathbb{C}^n$ which contains O. Let S be a closed subset of U such that $H_{2k+1}(S) = 0$. Then there are coordinates $z_1, \ldots, z_n$ on $\mathbb{C}^n$ and nieghborhoods, N_k of O in $\mathbb{C}^k$ and N_{n-k} of O in $\mathbb{C}^{n-k}$, such that $S \cap (N_k \times N_{n-k}) \xrightarrow{\Pi} N_k$ by $(z_1, \ldots, z_n) \longrightarrow (z_1, \ldots, z_k)$ is a proper mapping.

Proof: Suppose that for some $n-k$ dimensional subspace L of $\mathbb{C}^n$, $L \cap S$ is totally-disconnected. Choose coordinates so that L is $\{ z_1 = 0, \ldots, z_k = 0 \}$. This decomposes $\mathbb{C}^n$ into $\mathbb{C}^k \times \mathbb{C}^{n-k}$ with $L = \{O\} \times \mathbb{C}^{n-k}$. Since S is closed in U and $L \cap S$ is totally disconnected there must be a relatively compact neighborhood of O in $L \cap U$ whose boundary is disjoint from $L \cap S$. This neighborhood and boundary are of the form $\{O\} \times N_{n-k}$ and $\{O\} \times b_{n-k}$ where N_{n-k}

is a bounded neighborhood of O in $\mathbb{C}^{n-k}$ with boundary b_{n-k}. Thus $(\{O\} \times b_{n-k}) \cap S = \emptyset$ and $\{O\} \times \bar{N}_{n-k} \subset U$. Since b_{n-k} is compact and S is closed in U, if N_k is a small enough polydisk about O in $\mathbb{C}^k$ it will still be true that $(\bar{N}_k \times b_{n-k}) \cap S = \emptyset$ and $\bar{N}_k \times \bar{N}_{n-k} \subset U$. Therefore $S \cap (N_k \times N_{n-k})$ is closed in $N_k \times N_{n-k}$ and is at a positive distance from $\bar{N}_k \times b_{n-k}$.

Consequently, if Π: $S_n(N_k \times N_{n-k}) \longrightarrow N_k$ by projection and K is any compact set in N_k then $\Pi^{-1}(K)$ is closed in $S \cap (N_k \times N_{n-k})$ and is at a positive distance from both $\bar{N}_k \times b_{n-k}$ and $b_k \times \bar{N}_{n-k}$ (where b_k = boundary of N_k.) Therefore $\Pi^{-1}(K)$ is compact; and this shows that Π is a proper mapping.

It remains to locate an L such that $L \cap S$ is totally-disconnected. This can be done by a category argument - as follows.

The n-k dimensional subspaces of $\mathbb{C}^n$ form a complete metric space $\mathcal{S}$ by taking the Hausdorff metric on their intersections with the closed unit ball.

Now fix one coordinate system $z_1, \ldots, z_n$ for reference and let Z vary over all linear combinations of $z_1, \ldots, z_n$ with coefficients in $\mathbb{Q}(i)$. Let I vary over all intervals (a, b) in $\mathbb{R}$ whose end-points lie in $\mathbb{Q}$. Define $\mathcal{S}(I, Z) = \{ L \in \mathcal{S} : \operatorname{Re} Z(L \cap S) \supset I \}$. The relevant observation is that <u>if</u> $L \cap S$ <u>is not totally-disconnected then</u> L <u>belongs to some</u> (I, Z). (Take a Z such that Re Z is not constant on some non-trivial component of $L \cap S$.)

Therefore, it suffices to prove that the $\mathcal{S}(I, Z)$ do not exhaust $\mathcal{S}$. Since each $\mathcal{S}(I, Z)$ is clearly closed in $\mathcal{S}$ and the (I, Z) can be enumerated it is enough (by the Baire Category Theorem) to prove that each $\mathcal{S}(I, Z)$ has no interior in $\mathcal{S}$. But for that it will be enough to show that <u>if</u> $L_o \in \mathcal{S}(I, Z)$ and we express L_o as $\{ z_1^o = 0 , \ldots, z_k^o = 0 \}$ for suitable coordinates <u>then</u> the set of all $w = (w_1, \ldots, w_k)$ in $\mathbb{C}^k$ for which $L_w = \{ z_1^o + w_1 Z = 0, \ldots, z_k^o + w_k Z = 0 \}$ is an n-k space <u>in</u> $\mathcal{S}(I, Z)$ has no interior in $\mathbb{C}^k$.

Call this set W_o . Express I as (a, b) , and define a map $F : \{ p \in \mathbb{C}^n : \operatorname{Re} Z > a \} \longrightarrow \mathbb{C}^k \times \mathbb{R}$ by

$$p \longrightarrow \left(\frac{-z_1^o(p)}{Z(p)}, \ldots, \frac{-z_k^o(p)}{Z(p)}, \operatorname{Re} Z(p) \right) .$$

This is a Lipschitz map and, by assumption, $H_{2k+1}(S) = 0$. Therefore, by Property 2 , $F(S)$ has measure 0 in $\mathbb{C}^k \times \mathbb{R} = \mathbb{R}^{2k+1}$. But $F(S)$ contains $W_o \times I$. Therefore W_o cannot have any interior in $\mathbb{C}^k$.

Q. E. D.

CHAPTER IV

The Use of the Proper Mapping

We will now prove Theorem (C) and Theorem (D) .

THEOREM (C) . *Let* V_i *be a sequence of purely* k-*dimensional subvarieties of a domain* U *in* $\mathbb{C}^n$ *which converge to a (non-empty) limit set* V *in* U . *If* $\mathrm{Vol}_{2k}(V_i)$ *is uniformly bounded above then* V *is again a purely* k-*dimensional subvariety of* U .

Proof: If V is a variety it must be purely k-dimensional. For, $H_{2k+1}(V) = 0$ by the application of Proposition 3 ; and the local lower bound of Theorem (B) carries over in the limit, so that every non-empty open subset of V has positive 2k-measure.

To show that V is a subvariety of U I will prove that for each $p \in U$ there is a neighborhood $N \subset U$ such that $V \cap N$ is a subvariety of N . Since V is closed in U we may as well take $p \in V$; and translate so that p is the origin. Then , by Proposition 4 , there are coordinates $z_1, \ldots, z_n$ on $\mathbb{C}^n$ and neighborhoods, N_k of O in $\mathbb{C}^k$ and N_{n-k} of O in $\mathbb{C}^{n-k}$, such that if $N_n = N_k \times N_{n-k}$ and $\Pi: \mathbb{C}^n \longrightarrow \mathbb{C}^k$ by $(z_1, \ldots, z_n) \longrightarrow (z_1, \ldots, z_k)$, then the restriction of Π to $V \cap N_n$ maps $V \cap N_n$ properly to N_k . Also $\bar{N}_n \subset U$. Therefore, since $V_i \longrightarrow V$ in U , it follows directly that, for any relatively compact open subset $D_k \subset N_k$ and all i sufficiently large, the restriction of Π to $V_i \cap (D_k \times N_{n-k})$ is a proper mapping to D_k.

Let D_k be a polydisk and define $D = D_k \times N_{n-k}$. Let Π_i be the restriction of Π to $V_i \cap D$. Since $V_i \cap D$ is a purely k-dimensional variety and Π_i is a proper mapping to a polydisk in $\mathbb{C}^k$ it follows that

$$V_i \cap D \xrightarrow{\Pi i} D_k$$

is a finite-sheeted branched covering of D_k. (For a complete discussion of these matters read Chapter III , Section B of [3] , especially Theorem 21.) Let $s(i)$ be the associated sheet number . Then, if M is an upper bound for all the $\mathrm{Vol}_{2k}(V_i)$ we have $M > \mathrm{Vol}_{2k}(V_i \cap D) \geq s(i) \cdot \mathrm{Vol}_{2k}(D_k)$, so that the sheet numbers are also uniformly bounded above. Therefore we can extract a subsequence of the V_i (which will naturally have the same limit set V) all of whose members have the same sheet number s . Relabeling, call this subsequence V_i again.

The local analysis of a proper analytic mapping (as described in Chapter III of [3]) gives the following information .

For each i there is an open dense connected subset D_k^i of D_k such that

a) every bounded analytic function on D_k^i extends analytically over D_k (in fact, $D_k - D_k^i$ is a subvariety of D_k)

b) $\Pi_i^{-1}(D_k^i)$ is disjoint from the singular locus of V_i

c) $\Pi_i^{-1}(D_k^i) \xrightarrow{\Pi_i} D_k^i$ is a non-singular analytic even s-sheeted covering of D_k.

Also, for each $q \in V_i \cap D$, once G is a small enough neighborhood of $\Pi(q)$ in D_k then the sheet number of the restriction of Π_i to that component of $\Pi_i^{-1}(G)$ which contains q remains the same. Call this number $m_i(q)$, the multiplicity of Π_i at q. Then, for all $x \in D_k$,

$$\sum_{\Pi_i(q) = x} m_i(q) = s .$$

Therefore, since $\Pi : D \longrightarrow D_k$, we can define for each point u in D its <u>associates</u> $[A_1^i(u), \ldots, A_s^i(u)]$ as the <u>unordered</u> s-<u>tple of points</u> q <u>in</u> $V_i \cap D$ (<u>counted with multiplicity</u>) <u>for which</u> $\Pi_i(q) = \Pi(u)$.

Now I will complete the proof of Theorem (C) by constructing, for each point $v \in D - (V \cap D)$ a function h_v analytic on D such that $h_v = 0$ on $V \cap D$ but $h_v(v) \neq 0$.

First, fix v and pass to a subsequence of the V_i so that.

(after relabeling) for each $j = 1, \ldots, s$, $A^i_j(v)$ converges to some point $A^{\infty}_j(v)$ in $V \cap D$.

Next, let f_v be any bounded analytic function on D with the property that

$$f_v(v) \neq f_v(A^{\infty}_j(v)) \qquad j = 1, \ldots, s .$$

(For example, we could take f_v to be a suitable linear combination of $z_1, \ldots, z_n$.)

Then define functions h^i_v on D by

$$h^i_v(z) = \prod_{j=1}^{s} (f_v(z) - f_v(A^i_j(z))) .$$

These functions are polynomials of degree s in f_v whose coefficients are uniformly bounded (with respect to i) on D_k and are analytic on D^i_k . Therefore, these coefficients must be analytic everywhere on D_k , and $\{ h^i_v \}$ is a uniformly bounded sequence of analytic functions on D . Hence, some subsequence converges with local uniformity to an analytic function h_v on D . Evidently, $h_v(u) = 0$ if and only if $f_v(u)$ is the limit, for some j , of a certain subsequence of $\{ f_v(A^i_j(u)) \}$. Therefore, $h_v = 0$ on $V \cap D$ but $h_v(v) \neq 0$.

Q. E. D.

THEOREM (D). Let V be a purely k-dimensional subvariety of $\mathbb{C}^n$. Suppose there is a constant $K > 0$ such that $\mathrm{Vol}_{2k}(V \cap B(O; R)) \leq K \cdot R^{2k}$ for all $R > 0$. Then V is algebraic.

Proof: For each $R > 0$ let $\rho_R : \mathbb{C}^n \longrightarrow \mathbb{C}^n$ be scalar multiplication by R. Since all our coordinate changes and projections are linear they will commute with ρ_R. Define $V_R = \rho_{1/R}(V \cap B(O; R))$. Then each V_R is a purely k-dimensional subvariety of the unit ball $B(O; 1)$; and the condition that $\mathrm{Vol}_{2k}(V \cap B(O; R)) \leq K \cdot R^{2k}$ is equivalent to $\mathrm{Vol}_{2k}(V_R) \leq K$. Note also that every convergent subsequence of the family $\{V_R\}$ $(R \nearrow \infty)$ contains O in its limit set.

By the proof of Theorem (C), for some sequence $\{V_{R_i}\}$ (with $R_i \nearrow \infty$) there exist coordinates $z_1, \ldots, z_n$ on $\mathbb{C}^n$, a neighborhood $D = D_k \times D_{n-k}$ of O, and an integer $s > 0$ such that the projection $\Pi: \mathbb{C}^n \longrightarrow \mathbb{C}^k$ by $(z_1, \ldots, z_n) \longrightarrow (z_1, \ldots, z_k)$ induces, for each i, a proper s-sheeted branched covering

$$V_{R_i} \cap D \xrightarrow{\Pi_i} D_k .$$

Then the restriction of Π to each $\rho_{R_i}(V_{R_i} \cap D) = V \cap \rho_{R_i}(D)$ is again a proper s-sheeted branched covering

$$V \cap \rho_{R_i}(D) \longrightarrow \rho_{R_i}(D_k) .$$

Since the $V \cap \rho_{R_i}(D)$ exhaust V and the $\rho_{R_i}(D_k)$ exhaust $\mathbb{C}^k$ as $i \longrightarrow \infty$ it follows that

$$V \xrightarrow{\Pi|_V} \mathbb{C}^k$$

is also proper and s-sheeted.

This by itself will not insure that V is algebraic, but the proof can be completed in the following way. For each $p \in \mathbb{C}^n - V$ I will construct a polynomial vanishing on V but not at p.

To do this I first consider, for each point $z \in \mathbb{C}^n$, its associates $[A_1(z), \ldots, A_s(z)]$ with respect to $\Pi|_V : V \longrightarrow \mathbb{C}^k$ (determined by $A_j(z) \in V$ and $\Pi(A_j(z)) = \Pi(z)$.) Also, for each point $w \in D$ and each i, there are the associates $[A_1^i(w), \ldots, A_s^i(w)]$ with respect to $\Pi_i : V_{R_i} \cap D \longrightarrow D_k$.

There is a relation. Namely, for $z \in \mathbb{C}^n$ and all i so large that $\rho_{1/R_i}(z) \in D$,

$$(\ast)\quad [\rho_{1/R_i}(A_1(z)),\ldots,\rho_{1/R_i}(A_s(z))] = [A_1^i(\rho_{1/R_i}(z)),\ldots,A_s^i(\rho_{1/R_i}(z))]$$

(because ρ_{1/R_i} commutes with Π).

Now let

$$Z_p(z) = \sum_{j=1}^{n} c_j z_j$$

be any linear combination of the coordinates such that $Z_p(p) \neq Z_p(A_j(p))$ $j = 1,\ldots,s$. Define an analytic function h_p on $\mathbb{C}^n$ by

$$h_p(z) = \prod_{j=1}^{s} (Z_p(z) - Z_p(A_j(z))) .$$

Then

$$h_p(z) = \sum_{\mu=0}^{s} H_\mu(z_1,\ldots,z_k) \cdot (Z_p(z))^\mu$$

where each H_μ is analytic on $\mathbb{C}^k$. By the first formula for h_p it is clear that $h_p(p) \neq 0$ and $h_p = 0$ on V. Therefore, since Z_p is linear, it will suffice to prove that each H_μ is a polynomial, in fact of degree $\leq s-\mu$. By Cauchy's Estimate it is enough to show that, for all $(w_1,\ldots,w_k)$ in the neighborhood D_k of O in $\mathbb{C}^k$, $(R_i)^{\mu-s} \cdot H_\mu(R_i w_1,\ldots,R_i w_k)$ is bounded as $R_i \nearrow \infty$. To this end define analytic functions g_p^i on D by

$$g_p^i(w) = \prod_{j=1}^{s} (Z_p(w) - Z_p(A_j^i(w))) .$$

Then

$$g_p^i(w) = \sum_{\mu=0}^{s} G_\mu^i(w_1, \ldots, w_k) \cdot (Z_p(w))^\mu$$

and the functions G_μ^i are evidently uniformly bounded (with respect to i) on D_k . But since Z_p is linear the relation ($\ast\atop\ast$) yields $h_p(R_i w) = (R_i)^s \cdot g_p^i(w)$; so each $(R_i)^{\mu-s} \cdot H_\mu(R_i w_1, \ldots, R_i w_k) = G_\mu^i(w_1, \ldots, w_k)$ is bounded as $R_i \nearrow \infty$ and we are done.

Q.E.D.

Appendix

Let W be a subvariety of a domain U . Let V be a purely k-dimensional subvariety of U-W . The assertion of Theorem (F) is that <u>if</u> $\mathrm{Vol}_{2k}(V) < \infty$ <u>then</u> $\bar{V}$ (<u>closure in</u> U) <u>is a subvariety of</u> U . This can be proved by first demonstrating that $H_{2k}(\bar{V} \cap W) = 0$ and then proving Theorem (E) .

The proof that $H_{2k}(\bar{V} \cap W) = 0$ closely parallels that of Proposition 3 of Chapter III . However it is convenient to make the following reductions before proceeding. Firstly, we may localize and take U to be a ball. In that case W is cut out by global equations on U so we may replace W by some suitable $\{f = 0\}$. Finally, by embedding U in a higher-dimensional space via $(z_1, \ldots, z_n) \longrightarrow (z_1, \ldots, z_n, f(z_1, \ldots, z_n))$ we can arrange that W is a hyperplane.

Therefore, from now on we shall assume that W is the hyperplane $\{z_n = 0\}$. In proving that $H_{2k}(\bar{V} \cap W) = 0$, the assumption $\mathrm{Vol}_{2k}(V) < \infty$ plays the role of the upper bound. But besides that one needs a local lower bound of the following type.

(*) There is a constant $c(k,n) > 0$ such that if $p \in \bar{V}$ and $R > 0$ is so small that $B(p;R) \subset U$ then $\mathrm{Vol}_{2k}(V \cap B(p;R)) \geq c(k,n) \cdot R^{2k}$.

The case $k = 1$ is Theorem 2 of [1]. Here is a sketch of an argument of Bishop for the general case, by induction on k.

Remark. By integrating, it suffices to find $d(k,n) > 0$ so that $\mathrm{Vol}_{2k-1}(V \cap S(p;R)) \geq d(k,n) \cdot R^{2k-1}$ where $S(p;R) = \{p: \|p\| = R\}$. By translation and a change of scale we can restrict to the case $p = O$ and $R = 1$. Let $S = S(O;1)$.

LEMMA 1. *For each* $0 < r < 1$ *let* L_r *be the hyperplane* $\{z_1 = \sqrt{1-r^2}\}$ *in* $\mathbb{C}^n$ *and for each* $w = (w_1, \ldots, w_n) \in S$ *let* P_w *be the hyperplane* $\{\sum_{i=1}^{n} w_i z_i = 0\}$. *There is a constant* $c(n) > 0$ *such that*

$$(1) \quad \mathrm{Vol}_{2n-3}(L_r \cap S) \geq c(n) \int \mathrm{Vol}_{2n-5}(L_r \cap S \cap P_w)\, dS(w).$$

Proof: We have

$$(2) \quad \mathrm{Vol}_{2n-3}(L_r \cap S) = c_1(n) \cdot r^{2n-3}$$

where $c_1(n)$ is the volume of the unit $2n-3$ sphere.

$L_r \cap S \cap P_w$ is a $2n-5$ sphere with equations

$$z_1 = \sqrt{1 - r^2} \; , \quad \sum_{j=2}^{n} |z_j|^2 = r^2 \; , \quad \sum_{j=2}^{n} w_j z_j = -w_1 \cdot \sqrt{1 - r^2} \; .$$

Its radius is therefore $\leq r$ if $|w_1| \leq r$ and 0 otherwise. Hence, for $c_2(n)$ = the volume of the unit 2n-5 sphere, we have

$$\int \mathrm{Vol}_{2m-5}(L_r \cap S \cap P_w)\, dS(w) \;\leq\; c_2(n)\, r^{2n-5} \int_{|w_1| \leq r} dS(w) \; .$$

But for some $c_3(n) > 0$,

$$\int_{|w_1| \leq r} dS(w) \;\leq\; c_3(n) \cdot r^2 \; .$$

Then (1) holds with $c(n) = c_1(n)/(c_2(n) \cdot c_3(n))$.

DEFINITION. A 2k-1 <u>volume element</u> U of S is an open subset of $L \cap S$ where L is a complex linear variety in $\mathbb{C}^n$ of (complex) dimension k.

LEMMA 2. <u>If</u> U <u>is a</u> 2n-3 <u>dimensional volume element of</u> S <u>then</u>

(3) $\underline{\mathrm{Vol}}_{2n-3}(U) \geq c(n) \int \underline{\mathrm{Vol}}_{2n-5}(U \cap P_w)\, dS(w)$.

<u>Proof:</u> For a proper choice of coordinates $z_1, \ldots, z_n$ and

$r > 0$ L is the L_r of Lemma 1. We may decompose $L \cap S$ into a large number N of disjoint pieces which are nearly congruent under the unitary group, in such a way that U is very nearly the union of some M of these pieces. Then the two sides of (3) are very nearly M/N times the corresponding sides of (1). Thus (3) follows from (1) in the limit.

LEMMA 3. If U is a $2k-1$ dimensional volume element of S, $1 \leq k \leq n-1$, then

$$(4) \quad \underline{\mathrm{Vol}}_{2k-1}(U) \geq c(k+1) \int \underline{\mathrm{Vol}}_{2k-3}(U \cap P_w)\, dS(w) .$$

Proof: In an appropriate coordinate system $L = \{ z_1 = \sqrt{1-r^2}, z_{k+2} = 0, \ldots, z_n = 0 \}$. Therefore (4) reduces to (3) with n replaced by $k+1$.

Now, by the remark at the beginning of this discussion what we require is a constant $d(k,n) > 0$ such that $\mathrm{Vol}_{2k-1}(V \cap S) \geq d(k,n)$. By Theorem 2 of [1] we may take $d(1,n) = 2\pi$. Suppose we already have $d(k-1, n-1)$. Then approximate $V \cap S$ by the union of finitely many $2k-1$ dimensional volume elements $U_1, \ldots, U_m$ of S. By (4)

$$\mathrm{Vol}_{2k-1}(V\cap S) \approx \sum_{j=1}^{m} \mathrm{Vol}_{2k-1}(U_j) \geq$$

$$\sum_{j=1}^{m} c(k+1)\int \mathrm{Vol}_{2k-3}(U_j\cap P_w)\, dS(w) \approx$$

$$c(k+1)\int \mathrm{Vol}_{2k-3}(V\cap S\cap P_w)\, dS(w) \geq$$

$$c(k+1)\int d(k-1,\, n-1)\, dS(w) = c(k+1)\, d(k-1,\, n-1)\, \mathrm{Vol}_{2n-1}(S)\ .$$

Passing to the limit gives

$$\mathrm{Vol}_{2k-1}(V\cap S) \geq d(k,n)$$

with $d(k,n) = c(k+1)\, d(k-1,\, n-1)\, \mathrm{Vol}_{2n-1}(S)$.

This completes our sketch of Bishop's argument for (*) . It is now a simple matter to modify the proof of Proposition 3 to get $H_{2k}(\bar{V}\cap W) = 0$.

As for the proof of Theorem (E) I will make only the following comment. The condition $H_{2k}(\bar{V}\cap W)$ yields (similar to Proposition 4) , locally, a projection

$$\bar{V}\cap(N_k\times N_{n-k}) \xrightarrow{\ \Pi\ } N_k \subset \mathbb{C}^k$$

that is proper , and such that the image of $\bar{V} \cap W \cap (N_k \times N_{n-k})$ is closed and of measure zero in N_k . Bishop than makes a close analysis of this projection Π , using Rado's Theorem and some elementary properties of representing measures for uniform algebras, to get analytic equations for $\bar{V}$ on $N_k \times N_{n-k}$. This is Lemma 9 of [1] . The reader is now invited to turn to that paper.

Bibliography

1. E. Bishop, Conditions for the analyticity of certain sets, Mich. Math. Jour. 11(1964) 289-304.

2. H. Federer, Some Theorems on integral currents, Trans. A.M.S. 117(1965) 43-67.

3. R.C. Gunning and H. Rossi, "Analytic Functions of Several Complex Variables", Prentice-Hall, 1965.

4. T. Nishino, Sur les familles de surfaces analytiques, J. Math. Kyoto Univ. 1(1962) 357-377.

5. K. Oka, Note sur les familles de fonctions analytiques multiformes etc., J. Sci. Hiroshima Univ. A4(1934) 94-98.

6. K. Oka, Sur les fonctions analytiques de plusieurs variables, X. Une mode nouvelle engendrant les domaines pseudoconvexes, Jap. J. Math. 32 (1962) 1-12.

7. H. Rutishauser, Über Folgen und Scharen von analytischen und meromorphen Funktionen mehrerer Variablen, sowie von analytischen Abbildungen, Acta Math., 83(1950) 249-325.

8. W. Stoll, The growth of the area of a transcendental analytic set. I. and II, Math. Ann. 156 (1964) 47-78 and 144-170.

9. W. Stoll, Normal families of non-negative divisors, Math. Zeitschr. 84 (1964) 154-218.

10. G. Stolzenberg, A hull with no analytic structure, Jour. Math. and Mech., 12 (1963) 103-112.

11. W. Wirtinger, Eine Determinantenidentität und ihre Anwendung auf analytische Gebilde in Euclidischer und Hermitischer Massbestimmung, Monatsh. Math. Phys. 44 (1936) 343-365.

Also, as a further reference on volumes of analytic varieties there is

12. G. de Rham, On the area of complex manifolds, Seminar on several complex variables, Institute for Advanced Study, Princeton N. J. 1957.